Anastasia Shchurenko

Building an ecosystem and infrastructure for smart confectionery
Book 3

Anastasia Shchurenko

Building an ecosystem and infrastructure for smart confectionery Book 3

Innovative transformation of the ecosystem and infrastructure of a smart composite manufacturing facility

ScienciaScripts

Imprint

Cover image: www.ingimage.com

This book is a translation from the original published under ISBN 978-620-7-47296-3.

Publisher:
Sciencia Scripts
is a trademark of
Dodo Books Indian Ocean Ltd. and OmniScriptum S.R.L publishing group

120 High Road, East Finchley, London, N2 9ED, United Kingdom
Str. Armeneasca 28/1, office 1, Chisinau MD-2012, Republic of Moldova, Europe
Printed at: see last page
ISBN: 978-620-8-25623-4

Anastasia Shchurenko

Book - 3

Caption

Building a smart confectionery production ecosystem. Book 3

Subheading

Innovative transformation of the ecosystem and infrastructure of a smart facility for the production of composite products

Keywords:

Water regeneration, Water recycling, Electrochemical treatment of water in the regeneration process, Electrochemical process for heavy metal ions in water, Water flow through the electrode volume, Industrial wastewater treatment in confectionery, Filtration modules combined with ion exchange and other types of treatment, Aerodynamic turbo-flotation module, Electrochemical treatment reactor module, Ion exchange column module, Hydrodynamic mixing and homogenisation technology in confectionery production, Hydrodynamic mixing and homogenisation technology in confectionery production

Annotation

Innovative transformation of the characteristics of the ecosystem and infrastructure of the smart object of modern confectionery production into a developed combined supersystem with interconnected subsystems of on-line control in real time, including the ecosystem and infrastructure of integrated energy supply and water treatment modules;

Special complete smart technological equipment, tooling and tools for the production of ecosystem and infrastructure elements of smart confectionery production facilities and their general technical classical equivalents with complex gentrification of ecosystems of industrial zones and territories with revitalisation of architecture and general reconstruction of the environment of functioning of special technological equipment and tooling, maximally adapted to the ultra-clean confectionery production complex;

Technique, technology and design in the formation of construction systems and infrastructure inherent in smart plants and flow lines of confectionery production complexes and their innovative equivalents created in accordance with the development forecasts and created in related industries, allowing to perform the full classical working cycle as an example - in the regeneration and recycling of wastewater and adaptation of these technical solutions to the classical provisions of TRIZ and ARIZ, while initiating the processes of recreation, revitalisation and restoration.

Table of Contents

Introduction

Water treatment in production and storage facilities with special technological equipment and tooling to complete the base for smart confectionery production

Brief description of the features of the technology of electrochemical treatment of water for its subsequent use in the so-called smart confectionery, smart medical enterprises, biological and microbiological laboratories; options for using this technology in domestic conditions:

1. A brief introduction to the technology;

1.1. The proposed version of the technology is an electrochemical process of influence on water or aqueous solution, passing through a volume of electrodes connected to a source of direct current and separated by a neutral membrane;

1.2. The contacts that supply the electric potential to the electrode volume are made of non-metallic chemically resistant materials and are permeable throughout their volume to water and aqueous solutions;

1.3. The electrodes are permeable to water and aqueous solutions throughout their entire volume and are made of non-metallic, chemically resistant materials that are chemically compatible with the material of the electrical contacts;

1.4. The treatment process of water or aqueous solution takes place during its passage through the electrode volume, with the liquid passing through the electrode volume under the influence of gravitational forces in a directed upward flow;

1.5. The electrodes are arranged vertically and, the fluid inlet to the inner volume of the electrodes is at the bottom and the fluid outlet is at the top;

1.6. The time during which the liquid is in the inner volume of the electrodes is the time when the liquid is affected and the duration of this effect determines the level and depth of the effect on the performance and parameters of the treated water or aqueous solution;

1.7. The device for the proposed type of treatment of water and aqueous solutions, is an electrode cell with fluid input and output devices, connected to a source of electric potential;

1.8. Since this is a local application of the technology, the expected size and capacity of the plant is small; the main distinguishing parameter of the size range of such plants is their capacity, which varies from 50 to 250 litres per hour;

1.9. The units are versatile and their product can be the following types of specially treated water: - water with increased acidity level; - water with decreased acidity level; - alkaline water for use in various medical technologies and equipment; - acidic water for use in various medical technologies and equipment; - water purified from bacteria and microorganisms;

1.10. The units can also be used for electrochemical disinfection of laboratory and operating theatre effluents in hospitals and research centres;

2. Parameters and properties of water and aqueous solutions treated by the proposed method

2.1. The parameters of water with increased acidity are mainly determined by the Ph level; for initial tap water with Ph=7 units, after treatment it is possible to obtain water with Ph=3 units at about half of the initial volume;

2.2. The parameters of water with reduced acidity are also determined by the Ph level; by separating the initial water with neutral Ph=7 units it is possible to obtain after treatment water with pH=10 units at approximately half of the initial volume;

2.3. In water with an increased level of acidity as a rule in a combination of two types of influence - electrochemical due to high current density and electrochemical increase in the level of acidity are destroyed 100% of all bacteria and microorganisms;

2.4. The acidity level can be adjusted to the required values by changing the electric current modes and changing the flow rate and capacity of the device;

2.5. The parameters of items 2.1,- 2.4. - arise in a device with symmetrical electrodes and two inputs and outputs from the electrode cell;

2.6. Also contemplated is an embodiment of a device with asymmetrical electrodes, wherein in the case of an increased volume of the anode, an electrode with a positive electrical potential, an increase in acidity is obtained, and in the case of an increased volume of the cathode, an electrode with a negative electrical potential, a decrease in the acidity level in the entire volume of the treated water or aqueous solution is obtained;

2.7. If symmetrical electrodes are used, it is possible to recirculate one of the streams for re-treatment if necessary;

3. Examples of applications of water treated with the proposed technology:

3.1. Water with a high level of acidity; water used for sanitising rooms in hospitals; for sanitary antibacterial treatment of patients' bodies;

3.2. The water obtained after the reverse osmosis plant, treated to reduce the acidity level, can be used in autoclaves to produce steam without corrosive properties;

3.3. Water with a low level of acidity, can have a wide range of uses for burn wounds, for gargling, for sunburn prevention;

3.4. Aqueous, disinfected effluent may be discharged to the sewer system;

3.5. The disinfected water can be used for a variety of purposes;

4. Device Features:

4.1. The housing of the device is made of polymeric materials, mostly polyvinyl chloride;

4.2. The entire pipework design is made from standard components and can have many variations as required by the customer;

4.3. The electrodes are made of a composite material, carbon wool;

4.4. The contact devices are made of carbon fabric additionally saturated with carbon;

Electrochemical treatment of industrial wastewater in order to purify it from heavy metal ions and to enable the treated water to be reused repeatedly in a recirculation mode in a renovated production or office or storage facility

1. Treatment of industrial wastewater from heavy metals;

The main problem in modern industrial production is the necessity to use water for technological processes and to maintain the necessary level of water quality during the technological process; due to the fact that it is impossible to maintain the necessary level of water quality for a long time, the used water is removed from the process and new water with necessary properties and parameters for the process is introduced; due to the fact that the cost of water is constantly growing, the search for ways to reuse water resources is being conducted.

2. Information materials on global production associated with the generation of waste in the form of aqueous solutions containing metal ions;

Основными источниками загрязнения сточных вод ионами металлов являются:- промывные воды гальванических производств всех типов; - воды, использующиеся для охлаждения горячих металлических поверхностей; воды, использующиеся для промывки различных транспортных и технологических ёмкостей для транспортировки химических реагентов и материалов; - воды использующиеся в промывке после травления металлов; - воды, использующиеся в процессах нанесения металлизированных лаковых -красочных покрытий; воды, использующиеся в процессах горнодобывающих производс

In order to give an idea of the total amount of water that needs to be processed and treated, as an example, the water consumption for air-conditioning systems with water-cooled heat exchangers, for example in Israel, is about 150000000 cubic metres per year;

3. General classification of wastewater; place of restored premises in this classification;

Wastewater from industrial enterprises is divided into:

- waters that do not contain metals in ionic form;
-waters containing metals in ionic form in concentrations up to 1 gram per litre;

-waters containing metals in concentrations up to 5 grams per litre;

-waters containing metals in concentrations up to 10 grams per litre;

-waters containing metals in concentrations greater than 10 grams per litre;

in all of the above cases, metal ions may be in pure form and in the form of chemical complexes of metals and various non-metallic, including organic materials and compounds;

technologies for restored facilities are able to have a real impact on all types and classes of wastewater, including metal-free waters, in terms of acidity and alkalinity correction and turbo-flotation to separate or separate conductive and non-conductive components of the aqueous solution;

4. Classification of technical solutions for wastewater treatment; methods of wastewater treatment from metals;

About 16 different technological methods are known for the treatment of wastewater containing metals;

The most common method is the reagent treatment method, whereby caustic soda is added to the water to increase the alkalinity to a level where metal hydroxides begin to precipitate;

Artificial coagulation and flocculation methods can be disregarded as their economic criteria do not stand up to any criticism due to their high cost;

The electrocoagulation method can compete with the above mentioned reagent method in terms of cost and efficiency; this method has recently become more popular, especially for high-tech production facilities, but it cannot solve all the existing problems; the transition to the electrocoagulation system in the conditions of the enterprise operating the reagent method of purification involves significant capital expenditures, not solving in principle all the problems;
All known methods do not solve the problem of water purification for the purpose of its multiple use and, as a consequence, the problem of the need to reduce water consumption for technological needs;

5. What is the degree of popularity of existing water purification technologies;

As mentioned above, the most common technology is reagent-based wastewater treatment; recently, electrocoagulation technology is beginning to gain popularity;

6. An idea of, and a description of the boundaries of the scope of the proposed technologies; who is implementing similar technologies today? Who should the industrial and warehouse restorer take the market away from

These technologies are practically applied at all enterprises that use water for technological needs; there is a huge number of local integrator companies, which, using the general principles of treatment, depending on the conditions at the consumer, build treatment systems; large consumers, such as INTEL, resort to the services of large companies that concentrate the whole range of technologies for water treatment, but these large companies, such as VIVENDI, France, use the services of local integrator companies on site, using them in their own water treatment systems.

7. What market niches are vacant, for what reasons and how an industrial or warehouse restorer can occupy them;

The bulk of the technologies used in water treatment and purification processes are reagent-based; the niche without reagent-based technologies is practically free; even more free is the niche without reagent-based technologies that do not lead to the generation of difficult-to-dispose-of waste; practically today there are no such technologies on the market and 7 technological modules of industrial or warehouse restorers with all modifications and layout configurations, allow and give grounds to count on filling the niche with reagent-based technologies.

8. Who are today's main competitors to the technology of manufacturing and warehouse restorers;

At present, the technologies proposed by the restorers are not used elsewhere, due to the essential novelty of these technologies; once the elements of the technology proposed by the restorers are used, the most likely competitors will be large companies like ZENON in Canada and VIVENDI in France; especially new is the method proposed by the restorers of aerodynamic foaming and the resulting separation of conductive and non-conductive fractions in the flow of the treated water or aqueous solution; today, this trial is being carried out in the water or aqueous solution.

9. What are the technologies similar to the technology proposed by the restorers and what is the essential difference between the technology proposed by the restorers and the existing ones, what are the advantages and prospects for development;

The most advanced technologies for the treatment of metal-containing wastewater used at present are electrocoagulation technologies; for all its positive difference from older reagent methods, electrocoagulation does not solve all the acute issues facing water treatment processes and has a number of significant disadvantages, which are eliminated in the complex technology proposed by the restorers; the main advantage of the technology proposed by the restorers - one of which is the author of this publication - is the following - absence of necessity in the sedimentation process; - absence of toxic wastes requiring complicated and expensive methods of disposal; - insensitivity of the process to the level of acidity in the treated water and absence of necessity in correction of this acidity; - universality of electrochemical modules - reactors, allowing their inexpensive and simple modification for specific conditions at the consumer, - customisation, while preserving the basic technological and constructive principles; - possibility of wide advanced unification of modules - reactors; - possibility of in-depth unification of modules - reactors. -

10. Description of the proposed device; size range; performance; possible applications;

A complex of technological modules is offered, each of which is an independent subject of technological equipment, capable of functioning as a part of any technological line for water and aqueous solutions treatment; in total, 7 types of technological modules are offered, each having a certain number of modifications and size categories;
The proposed modules have the following technological characteristics: modules for accumulation of liquid before and after treatment; these modules are cylindrical tanks with a conical base and include level, temperature, acidity or alkalinity sensors, agitators

and other standard instruments and devices, depending on the configuration of the process line;

Filtration modules combined with ion exchange and other types of treatment; these modules have two basic designs, one for inlet filtration and one for final filtration; these modules vary in modifications and size categories and are original technology of the author's company;

Aerodynamic turbo-flotation module; utilises an aerodynamic foam generator to generate foam in the dynamic upward flow of the treated fluid; the module and all its technological and design elements are original technology of the author's company;

The module is an electrochemical treatment reactor; the main technological module of the complex; it has 12 versions for different applications; it is an original technology of the company of the author of the publication;

Module - column for ion-exchange treatment; the module is able to work with synthetic and natural ion-exchange materials, in combination or without combinations with other materials; the module is an original technology of the company of the author of the publication;

All the modules listed above have several gradations of different capacities - 500 litres per hour; 750 litres per hour; 1000 litres per hour; 1250 litres per hour; 1500 litres per hour; the above 7 modules can be used to make up technological lines for regeneration and recycling of water in various technological processes; the configuration of the lines, thanks to the modular design, has a high level of layout and technological flexibility; thanks to the complete autonomy of the modules, they can be integrated into existing water treatment lines; the same is possible in combinations of the modules.

11. Plant Maintenance;

The modules proposed by the author of the publication are designed in such a way that maintenance and operational monitoring operations are minimised; the modules can be controlled in fully automatic mode or in any other less complex variants; in all modules, routine maintenance is reduced to the replacement of disposable technological elements; the time required for these operations is minimised and such operations do not require special operator training; there is a possibility for remote monitoring, which is very important.

12. Plant fabrication; cost of production; marketing options; cost of options; prices to consumers;

To manufacture modules do not require special materials, components and technologies; any medium-sized engineering enterprise is able to organise the manufacture and assembly of modules; given the fact that the structure of all equipment proposed by the author of the publication is modular, and each module is functionally complete and autonomous apparatus, serial production increases, which positively affects the cost of production; due to the high degree of unification of module designs, there is a possibility of manufacturing different

The author of this publication considers it necessary to describe the most important elements of the described technologies

Electrolytic reactor for producing an aqueous solution of oxygen and hydrogen by electrolysis;

1. Equipment
2. Methodology
3. Mounting on the fuel system
4. Distinctive design features
5. Distinctive technological features
6. Advantages derived from design features
7. Advantages derived from technological features
8. Comparative analysis with known technical solutions aimed at realising a similar task
9. Potential economic performance
10. Initial drafting of the claims

10.1. A device for energetically modifying a fuel mixture before it is fed into a combustion chamber, comprising:

- Fuel tank with fuel pump connected to the fuel line;

- A device mounted on the fuel line for sequential hydraulic and pneumatic treatment and vortex mixing of fuel mixture components with oxidising compressed gas, with the output of the device connected to the inlet of the fuel injector;

- Systems for supplying additional fuel components to an area of said fixture in which reduced pressure is created;

- Systems for supplying oxidising compressed gas to the area of said fixture, whereby it is used to create a reduced pressure;

- A water solution storage tank connected at the inlet to the condensate outlet from the air conditioning system and at the outlet to the inlet of the electrolytic reactor;

- An electrolytic vertical reactor having at least one pair of electrodes connected to a source of direct electric current, one inlet and one outlet at the bottom and two outlets at the top;

- Systems for storing and feeding gaseous electrolysis products into a system for supplying additional fuel components to the fixture area where reduced pressure is created;

- A system for discharging aqueous solution from the space between the electrodes into an outlet located in the lower part of an electrolytic vertical reactor, wherein at least one of the electrodes of said reactor is cylindrical, made of a non-metallic material permeable to aqueous solution and installed vertically and concentric to the outlet channel in the lower part of the vertical reactor, connected to one of the outlet channels in the upper part of the vertical reactor, which also has a free outlet to the atmosphere;

10.2. A method for energetic modification of a fuel mixture, before its supply to a combustion chamber by transforming the fuel mixture flow, forming a low-pressure zone along the flow of the mixture to the combustion chamber and introducing an additional fuel component into this zone, and subsequently mixing the fuel components with a compressed air flow and forming multilayer bubbles in the fuel mixture flow, and subsequently foam from the fuel mixture, comprising:

- preparation of aqueous solution, for electrolytic preparation of gaseous oxygen and hydrogen;
- introducing the said aqueous solution into the interelectrode space between the coaxial electrodes, the said space being in the form of an ascending spiral;
- applying electrical potential to the electrodes;
- ensuring the movement of aqueous solution along the ascending spiral in the interelectrode space;
- electrolytic decomposition of aqueous solution with release of gaseous oxygen and hydrogen and movement of gases in an upward spiral in the flow of aqueous solution;
- overflow of the liquid fraction of the aqueous solution into the outlet system and outlet of the gaseous mixture into the piping system associated with the depressurised zone;
- drawing the gases into the low-pressure zone and mixing them with the fuel mixture flow;

Comments in addition to the material presented; question - Electrochemistry,

Currently, electrochemical technologies are widely used in industry for electrocoagulation of aqueous solutions resulting from economic activities. The most difficult problem is the extraction or purification from heavy metals, which are in aqueous solutions in ionic form, or in the form of various water-soluble compounds. There are a number of electrocoagulation technologies that are more or less effective, but all of them have a major disadvantage - the sludge remaining after treatment must be disposed of. This is an extremely difficult problem, which negates the positive effect that occurs when comparing electrocoagulation technologies with technologies of reagent treatment of aqueous solutions.

The second problem, which is no less severe, is the problem of sedimentation after the electrocoagulation process. This process requires a large production area, has not changed to date and is very inefficient. It is not suitable for modern technologies for a number of well-known reasons, which makes it necessary to move this process outside of high-tech production areas.

Regeneration of process water and return of this water to the production process after regeneration is not possible with the old methods of water treatment.

There is another technology that allows electrochemical deposition of heavy metals on the cathodes of electrochemical cells using galvanic deposition technologies. In this process, the metal is obtained in solid form and it does not require the disposal of regeneration products after water treatment. This positive quality does not find a place for use in industry due to several significant reasons:

- this process can only proceed with initial concentrations of heavy metals within the range of at least 5 grams per litre;
- this process cannot purify water to concentrations that allow its reuse;
- this process depends on initial concentrations of mineral oil and total organics in water, i.e. their presence reduces the speed and efficiency of the process and does not allow its application in conditions of automated production;
- maintenance of this type of equipment is difficult and requires stopping the main process equipment when performing maintenance operations.

Thus, in order to improve the efficiency of the said process, it is envisaged that:

- lower the initial concentration limit of heavy metals in treated water to 0.01 grams per litre;
- lower the residue limit to 0.1 milligrams per litre;
- Provide a disposable option for the use of electrodes;
- To provide an efficient and compact process mechanism for flotation ;
- to provide in the technology two technological processes, in which the process of metal extraction is carried out on the cathode, and on the anode - the process of neutralisation and disinfection of aqueous solutions;
- to provide simultaneously with metal recovery the process of reduction and oxidation of toxic components in aqueous solutions;

Electrochemistry, comments to follow

A new technology for extraction of heavy metals from process water containing electrically active components in small concentrations within the range (for initial concentrations) from 5 grams per litre to a minimum of 0.01 grams per litre and bringing to residual concentrations of less than 0.1 milligrams per litre is proposed.

The design of the equipment uses electrolysers with flow electrodes with highly developed reaction surface, allowing to intensify the electrolysis process significantly. In the project these electrolysers are called Electrochemical Reactor (ECR).
Flow-through volumetric porous electrodes are the design element that makes it possible

to achieve the intended performance parameters.

The electrolysis process , is also one of the main elements of the new technology and it proceeds as follows :

- The treated aqueous solution is pumped through the pores of the electrodes, concentrating the metal in the cathode volume, which can then be produced as ingot, foil or concentrate by pyrometallurgical, electrochemical or chemical methods;

- electrode chambers of the electrochemical reactor are separated by an ion-exchange or neutral membrane, which allows their autonomous feeding with solutions and efficient use of two electrode processes, for example, cathodic for metal extraction and anodic for solution neutralisation.

As materials for electrode construction it is proposed to use non-woven composite fabric for the main, metal-absorbing volume (carbon-graphite material) and, which is very important and has a pioneer status as an invention, to apply elastic permeable contacts made of carbon-carbon composite, in the form of carbon composite fabric made by pyrolysis.

At consumption of maximum 10 amperes of current per 1 square decimetre of the conventional area of the electrode and at a voltage of 6-12 volts, the proposed technology surpasses all known technologies by 100 times in terms of speed and quality of metal extraction, with the equivalent size of the electrochemical reactor.

Assumed technical characterisation :

DC load, Amps, - REACTOR);	150 A (PER ELECTROCHEMICAL
Voltage,-	6-12 volts;
Number of anode chambers,-	1;
Number of cathode chambers,-	1 ;
Maximum number	
Metal deposited in the cathode	
Electrolyser, -	8kg ;
Metal recovery percentage, %,-	99,5 ;
Productivity,-	0.3 cubic metres per hour ;

COMPOSITE DAIRY FOOD PRODUCTS AS A CONSEQUENCE OF INTEGRATIVE INNOVATIVE TECHNOLOGICAL PROCESS

It is a common myth that butter is harmful because of its high cholesterol content.

But cholesterol, as such, is not dangerous, on the contrary - necessary for the body, dangerous metabolic disorders as a result of which cholesterol begins to deposit on the walls of blood vessels.

There is no way that consuming butter can cause such serious consequences.

It is good for eyesight, ophthalmologists even call it prolonging the youthfulness of the eyes, and butter also contains fatty acids and numerous vitamins.

Nutritionists recommend that healthy people eat 10g of fresh butter a day - this amount is within the calorie allowance.

Butter should be consumed in its natural form - frying loses more than 50% of nutrients, vitamins are destroyed, and useful unsaturated fatty acids turn into harmful saturated ones.

Butter is not only nutritious but also beneficial for skin, hair, eyesight, bone and muscle tissue. It is rich in vitamins, calcium, and contains phospholipids, which are essential for building cells, especially nerve cells.

In addition, butter contains essential amino acids - substances necessary for the normal functioning of the body, which come only with food.

So there is no doubt about the usefulness of butter. We just have to remember that when we melt it in a frying pan, there are no vitamins left in it. That is why it is better to put a piece of butter in already cooked food

Of course, butter also contains cholesterol. But it becomes harmful only with immoderate consumption of fatty foods, and in small doses is even necessary, because it participates in the construction of the walls of blood vessels.

So one or two fresh butter sandwiches a day cannot harm our health.

There are 20 known basic principles and attributes that reveal and explain the unique properties of butter:

1. Butter is rich in the most easily absorbed form of Vitamin A, which is essential for normal thyroid and adrenal gland function

2. Butter contains lauric acid (lauric), the presence of which is extremely important from the point of view of prevention and avoidance of infections, and in the accumulation of biological mechanisms of resistance to infection

3. Butter contains lecithin, the basis for stabilising the biological mechanism for the formation of cholesterol metabolism.

4. Butter contains antioxidants that protect the body from spontaneous and radical biological degradation

5. Butter has antioxidants in its composition that protect the body's circulatory system from weakening and thinning of the arterial walls.

6. Butter is an important and effective source of extremely beneficial Vitamins E and K.

7. Butter is a very rich and unique biological source of the body's vital mineral selenium.

8. The saturated animal fats in butter have strong properties to influence the inhibition of the formation of various tumours and cancers.

9. Butter contains amino acids, which are powerful agents in preventing the formation of cancerous tumours, stimulating the development of muscles, and a stimulant for the restoration and inviolability of the immune system.

10. Vitamin D found in oil significantly affects the calcium absorption process.

11. Butter protects against decay of tooth enamel.

12. Butter is the only source of a factor that protects against joint hardening.

13. The factor of fatty acids in butter also prevents calcification of arteries, streams, and hardening of glands.

14. Butter is a source of Activator X, which helps your body absorb beneficial micronutrients.

15. Butter contains iodine in a very absorbable form.

16. Butter can promote reproductive functions in the body in women.

17. Butter is a source of quick energy.

18. Positive Cholesterol found in milk fat is very important for children's brain development and nervous system development.

19. Butter contains Arachidonic acid (AA) , which plays an important role in brain

function and is a vital component of cells.

20. Butter protects the body from gastrointestinal infections at a very young age or in old age

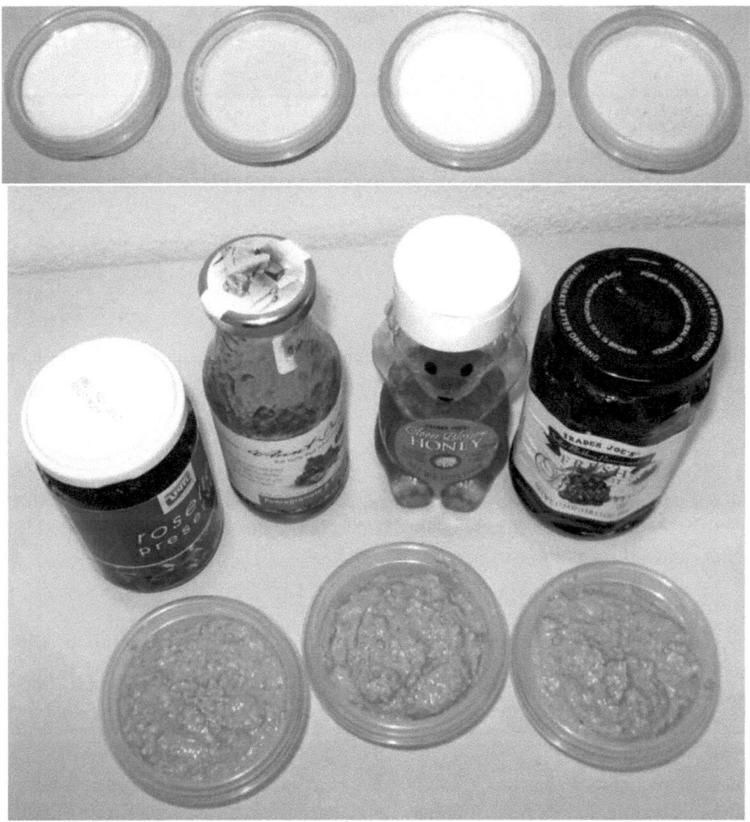

The author of this publication and technology developer offers you an innovative complex group technology for the production of especially pure food composites for the following product groups:

Homogenised Meta-Pure food composition - cream

Homogenised Meta-Pure food composition - butter

Depending on the version and strategy of introduction of these products on the market, the company - partner of the author of this publication has the rights of the patent owner and can grant licences for the right to use the following universal technologies, including nano-technologies with elements of artificial intelligence and artificial neural networks: Technology of active non-contact on-line control in real time mode of the composition of components of innovative products and their concentration in liquids, including cow's milk, as well as control of the number of somatic cells and twenty other basic components and parameters of cow's milk;

Technology of reagent-free washing, sterilisation and disinfection of process pipelines and containers in production lines for dairy products and confectionery;

Technology of three-dimensional dynamic homogeneous fermentation of dairy products and confectionery components;

Technology of vortex micro-aeration of dairy products and components of confectionery products;

Technology of hydro-dynamic harmonic pulsating shock kinetic impact on the volume of liquid (including milk, cream and sour cream and liquid components of confectionery products);

Food and confectionery products, as well as pharmaceuticals and, to a certain extent, cosmetics, represent a particularly quality-controlled sector of the innovation field of inventors and advanced industries;

Above all, this field is characterised by the need for new innovative products and technologies to fully comply with health standards and increasingly strict and stringent environmental regulations;

It must be recognised that this technological field is very much influenced by the traditions, climatic and cultural peculiarities characteristic of the region in which the innovative confectionery product is offered to the consumer;

But it is also important to recognise that the most successful local innovations and inventions, which have applications not only in the area limited to food, but also in areas of general technology, can serve as effective pushes in the progressive development of integrative innovation technologies.

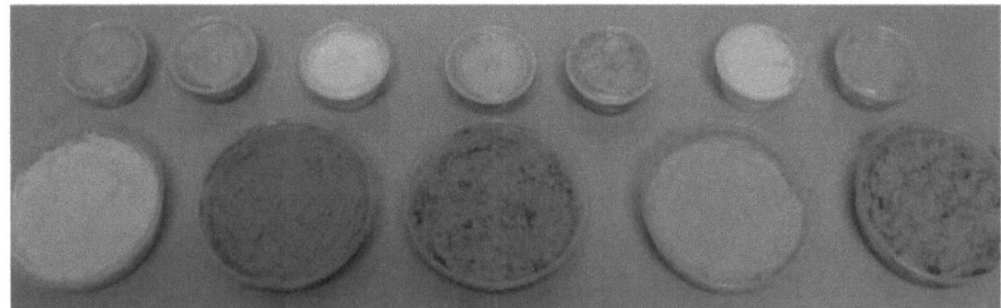

HONEY
CROWN
Fresh

365
ORGANIC
Parsley
Nutrition Facts

Presumably one such technology is brought to our attention;

New technology of hydrodynamic mixing of various liquid components, with simultaneous knockdown in dynamic, periodically repeating, developed kinetic mode with regeneration of accumulated kinetic energy and its consecutive transfer into the process

The proposed innovative technology is a process of dynamic influence on the liquid or partially consistent components to be mixed and knocked down;

The most researched, and often in different variants used in this process, are components from various dairy products;

The company-developer (company name) conducts this kind of research and development of special technological equipment;

The purpose of the above research, development and design work, as well as the formation of nutritional algorithms and programmes, is to find new ways for the production of composite, compound or homogeneously combined food products, which have no or minimised fat content and, on the contrary, increased concentration of vitamins and biologically active elements;

Additional purpose of realisation of the offered process, is possibility without heat treatment and without other kinds of influence on the product, as it is already told, mainly dairy or lactic acid, but also can be in the form of components of confectionery products, to receive a mixture or whipped mixture of components which in usual conditions either do not mix at all, or poorly mix or are not stable;

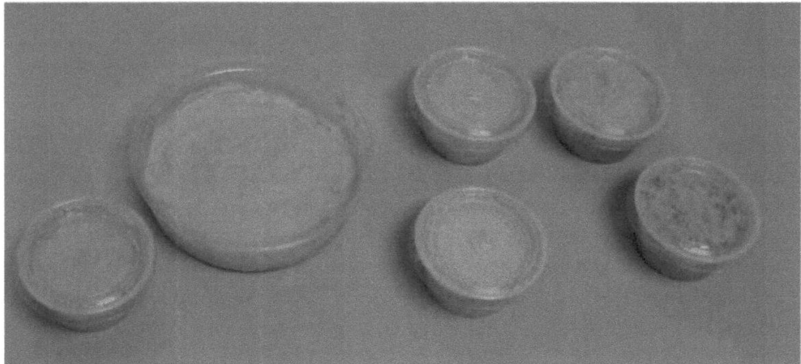

Principle differences between the butter churning process and known processes for the same purpose :

- knocking down is carried out in a flask having the shape of a cylinder, closed at the ends with hemispherical lids having conical holes in the axis;

- the whole structure consisting of a cylindrical jacket with conical sharpenings on its ends and two hemispherical lids having conical chamfers on the base and coaxial to them conical holes on the axis, assembled and sealed by inserting into the conical holes of the lids, previously aligned with the cylindrical jacket by conical sharpenings on its ends and conical chamfers on the base of the lids, - clamping, orienting, fixing and sealing the internal volume of the said structure, fixing clamping devices installed

- Due to these design features, the churning process gains, along with significant differences, significant advantages in the quality of the obtained product and the possibilities of the technological range of the oil churning process;
- additional possibilities in the process realised on the proposed device:
- The volume in which the oil is churned is collapsible;

- in the process of knocking down there is a possibility of authorised penetration into the internal volume, introduction of the components of the mixture to be knocked down, removal of waste or part of the components from the working volume;

- conical holes of hemispherical lids contribute to a more complete breakdown of the mixture when hitting the inner hemisphere of the lids, i.e. they fulfil a number of combined functions, such as activating, combining and orienting, sealing, fixing, and, most importantly, they fulfil functions of

the process airlock on both sides of the structure, both outlet and inlet, depending on the position of the flask;

- the specified design allows to have several sets of flasks on one device for beating, which allows, with a relatively short duration of the process of beating, to change the composition of components for beating, including the nature of the lactic acid matrix, the degree of its fat content and other parameters;

- hydrodynamic parameters of the knocking down process, in the form of flask elements, in the presence of a spherical depression with an activating cone in the centre at the end of each linear displacement of the knocked down mass, allow to significantly increase the efficiency of the knocking down process, which is explained by the fact that the volume of the knocked down mass at each blow to the lid, occurring after half a turn of the leash frame, is transformed, passing within a very short period of time from cylindrical to spherical form, i.e. there is a process of volumetric deformation at the same time.

- the described character of influence on the mass to be knocked down, due to the fact that there is a kinetic variety of types of forced changes in the volume and geometric shape of the mass to be knocked down, under the influence of many force factors caused by the combination and concentration of movements and loads caused both by the shape of the flask geometry and the character of movements caused by the kinematics of the oil knocking apparatus system.

From the presented models of hemispherical lids, it can be seen that :

- their design is technologically advanced;
- they are highly durable;
- they are easy to maintain;
- they're hygienic;
- They can be made from a wide variety of materials;
- for the same cylindrical jacket design it is possible to have different variants of covers, differing in the steepness of the sphere and the depth of the conical hole along the axis, while maintaining the identity of the jointing elements of the design.

Butter churner - description of application principles

The butter churning appliance is designed for efficient churning of butter from lactic acid raw materials, mainly sour cream and mainly with low fat content. Preliminary tests of the device have shown the possibility of efficient churning not only from pure sour cream, but also from sour cream with various additives, mainly of vegetable origin.

Such additives may include the following materials:

- carrot juice;
- carrot puree;
- tomato juice;
- tomato paste;
- cucumber brine;
- cucumber puree;
- mashed avocado;
- mashed kiwi;
- mashed sweet peppers;
- various courgette purees;

- various fruit purees, - apple, plum, peach, pear, apricot, pineapple. Puree from tropical fruits and vegetables;

- various combinations of the above additives , and in combination with dill, garlic, onions, lemons;

- with prune pulp;

- jam from black mountain ash; sea buckthorn and other medicinal berry plants;

- various purees of strawberries, strawberries, blueberries; blackcurrants; raspberries; mulberries; calamus; gooseberries and other berries;

The above list can be continued and various vegetables and fruits from local geographical areas can be introduced, also in various combinations and combinations with known products.

It is possible to modify the oil by introducing vegetable fats during the mixing process in order to change the balance of vegetable and animal fats in the oil;

In order to extend the scope of application of the butter churner, the design of the butter churner allows it to be flexibly adapted to produce butter with a wide variety of additives and in a wide variety of volumes.

Versatility, as the ability to quickly change to churning butter with other components, is one of the main operational advantages of the machine and the churning process itself.

An apparatus of a mixer flask having three constituent elements, of which the original and maximally influencing the results of mixing and obtaining commercial products with different properties are:

- jacket of the mixer flask, having conical recesses on the flanges, for sealing of the internal volume and simultaneously for centring and volumetric orientation of the flask geometrical parameters in three-dimensional space, allowing to keep all geometrical spatial relations, including relations of the apparatus axis positions, in the process of mixing;

- conventionally upper hemispherical lid having two conical surfaces located on one axis, intended for centring, orientation and sealing of the inner volume of the flask, and allowing to introduce all necessary components of the initial composition for oil knocking into the upper one of them, taking axial force for fixation and local sealing of the flask from the fixer with axial overlapping hole, when the flask structure is assembled and sealed.

- conditionally bottom hemispherical lid, having three conical surfaces located on one axis, designed for centring, orientation, sealing and freeing from the waste from the knocking down of the internal volume of the flask, and allowing the waste generated from the oil knocking down process to be discharged into the lower one of them, taking the axial force for fixing and local sealing of the flask, from the fixer with a ball valve, when the flask structure is assembled and sealed.

The process of beating butter without additives consists of the following operations :

After washing from the previous mixing cycle, the flask and the upper and lower hemispherical lids are preliminary aligned on conical surfaces, placed in the bracket and fixators begin to compress along the axis, until complete alignment of conical surfaces - two in the upper lid and jacket of the mixer flask and two in the lower lid and jacket of the mixer flask. After that, open the sealing plug in the upper fixer and pour the required volume of sour cream into the inner volume of the flask. Then close the stopper in the lock and begin to rotate the frame with the flask. The sour cream moves along the axis of the inner flask cavity, making one complete movement with a bump in the bottom of the lid for every half turn of rotation of the frame. After completion of the churning cycle, the system is stopped and the frame is rotated and the flask is placed in a position where the bottom lid is at the bottom, the ball valve is opened and the waste from churning the butter is poured out of the inner cavity of the flask. After that, release the clips and remove the flask with the lids from the frame; then remove both lids and the beaten oil is removed from the inner cylindrical cavity of the jacket of the mixer flask. After that the three constituent elements of the flask are washed and a replacement set is installed in their place.

In case it is necessary to introduce additives or a combination of additives into the oil, all components are introduced according to the above described scheme. If it is necessary to introduce the components sequentially or at different time intervals, the frame must be stopped and the locking plug must be opened before introducing each of the components. The possibility of flexible introduction of components into the process, with orientation either by time or by the degree of oil readiness, or by any technological reaction, e.g. in case of temperature changes or chemical-technological features, significantly extends the technical and technological range of the oil churner.

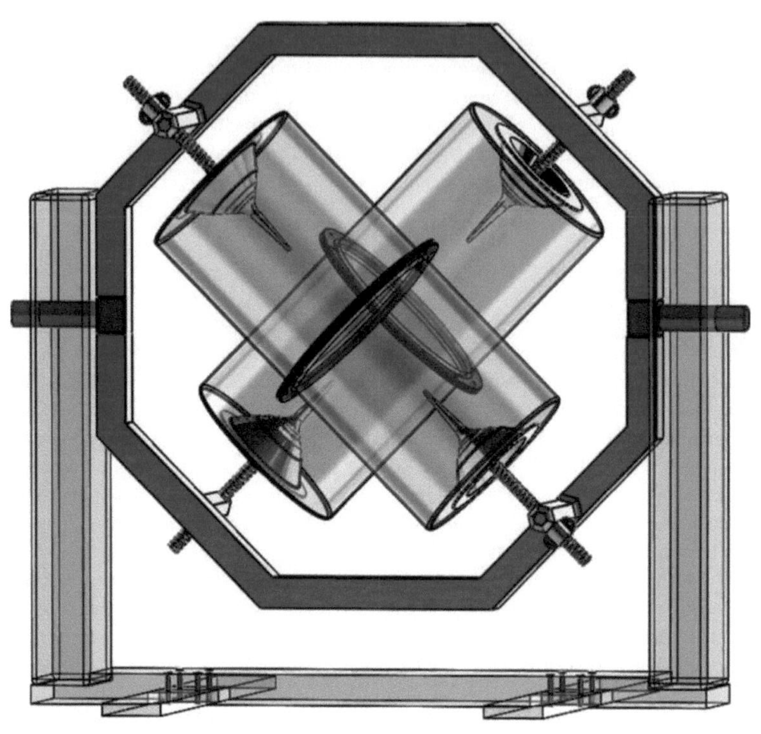

Appendix 01

There are 20 basic principles and features that reveal and explain the unique properties of butter:

1. Butter is rich in the most easily absorbed form of vitamin A by the body necessary for normal functioning of the thyroid and adrenal glands

2. Butter Contains laurel acid (lauric), the presence of which is extremely important in terms of prevention of infections and the accumulation of biological mechanisms of resistance to infection

3. Butter contains lecithin, the basis for the stabilization of the biological mechanism for the formation of cholesterol metabolism.

5. Butter contains antioxidants that protect the body from the spontaneous and radical biological degradation

6. Butter has in its composition of antioxidants that protect the body's circulatory system from weakening and thinning of the walls of the arteries.

7. Butter is an important and effective source of extremely useful vitamins E and K.

8. Butter is a very rich and unique biological sources vital for the body's mineral selenium.

9. Saturated animal fats in butter have strong influence on the properties to prevent the formation of a variety of tumours and cancer.

10. butter contains amino acids, which are a powerful agent prevents the formation of cancerous tumours, stimulates the development of muscles and stimulating factor for the restoration of immunity and inviolability.

11. Vitamin D is found in the oil significantly affects the absorption of calcium.

12. Butter protects against decay of the tooth enamel.

13. Butter is the only source factor that protects against hardening of the joints.

14. Factor presence of fatty acids in butter also prevents calcification of the arteries, streams, and hardening of glands.

15. Butter is a source of Activator X, which helps your body absorb the beneficial minerals.

16. Butter contains iodine in a very absorbed form.

17. Butter can promote reproductive functions in the body in women.

18. Butter is a source of quick energy.

19. Positive Cholesterol that soldierships in milk fat lochan important for the development of a child's brain and nervous system development.

20. Butter Containing Arachidonic acid (AA), which plays an important role in brain function, and is a vital component of cells.

21. Butter protects the body from the gastrointestinal infections in the very young or the elderly

LIST OF USED LITERATURE, PATENT AND LICENCE MATERIALS

APPENDIX 1

United States Patent **10,760,484.**
Alecu **September 1, 2020**

Multi-engine aircraft power plant with heat recuperation

Abstract

Multi-engine aircraft power plants and associated operating methods are disclosed. An exemplary ***multi-engine*** power plant comprises a first turboshaft ***engine*** and a second turboshaft ***engine*** configured to drive a common load such as a ***rotary*** wing of an aircraft; and a heat exchanger in thermal communication with an exhaust gas of the first turboshaft ***engine*** and in thermal communication with pre-combustion air of the second turboshaft ***engine***. The heat exchanger is configured to permit heat transfer from the exhaust gas of the first turboshaft ***engine*** to the pre-combustion air of the second turboshaft ***engine***.

APPENDIX 2

United States Patent **10,760,471.**
Yamamoto , et al. **September 1, 2020**

Auxiliary machine-driving device for vehicle

Abstract

An auxiliary machine-driving device is provided for a vehicle. The auxiliary machine-driving device has a first roller, a second roller, a third roller, a fourth roller and a fifth roller. The first roller rotates integrally with a ***rotary*** shaft of an ***engine***. The second roller rotates integrally with a ***rotary*** shaft of a motor/generator. The third roller rotates integrally with a ***rotary*** shaft of an auxiliary machine. The fourth roller is provided between the first roller and the second roller. The fifth roller that always contacts the second roller and the third roller. The actuator switches the fourth roller between a contact state with the first and second rollers and a separation state from the first and second rollers.

ANNEX 3

United States Patent **RE48,173**
Takahashi , et al. **August 25, 2020**

Vehicle driving device

Abstract

A vehicle driving device that includes a ***rotary*** electrical machine that functions as a driving force source of a wheel along with an internal combustion ***engine;*** a transmission device that is disposed .[.side by side with respect.]. .Iadd.next .Iaddend.to the ***rotary*** electrical machine in an axial direction, the axial direction being a direction in which a rotation axis .[.centre.]. of the ***rotary*** electrical machine extends; a case including a first case section that accommodates the ***rotary*** electrical machine and a second case section that accommodates the transmission device; an inverter device that controls the ***rotary*** electrical machine; and a wiring .[.member.]. .Iadd.structure .Iaddend.that connects the ***rotary*** electrical machine and the inverter device.

APPENDIX 4

United States Patent **10,753,313.**
Okagawa , et al. **August 25, 2020**

Rotating part supporting structure for engine

Abstract

Disclosed herein is a support structure for a ***rotary*** part of an ***engine*** (2). The support structure includes crank journal bearing metals (14) and crank pin bearing metals (24, 124). Each crank journal bearing metal (14) includes chamfers (16) and crowned portions (17). On the other hand, each crank pin bearing metal (24, 124) includes chamfers (26) and no crowned portion or crowned portions (127) on its inner peripheral surface. The crowned portions (127) are inclined at a smaller angle than the crowned portions (17).

APPENDIX 5

United States Patent	**10,731,548.**
Yoeda	**August 4, 2020**

Mechanical supercharging system

Abstract

A mechanical supercharging system includes a stepped transmission that connects a crankshaft of an internal combustion ***engine*** with driving wheels, a centrifugal supercharger including a ***rotary*** drive shaft connected to the crankshaft, a variable speed ratio device that changes a speed ratio of the ***rotary*** drive shaft to the crankshaft, the variable speed ratio device being provided between the crankshaft and the ***rotary*** drive shaft; and a control device configured to control the speed ratio. The control device increases the speed ratio during the upshift operation more than the speed ratio before start of the upshift operation.

ANNEX 6

United States Patent	**10,697,366.**
Liu , et al.	**June 30, 2020**

Rotary engine with axially directly connected compression and power cylinders

Abstract

A ***rotary engine*** with axially directly connected compression and power cylinders is disclosed, which includes a compression cylinder, a power cylinder, an intermediate cylinder wall located between the compression and the power cylinder to serve as a common inner-end wall of the two cylinders, and a combustion chamber unit fixed to a circumferential surface of the intermediate cylinder wall, so that the ***rotary engine*** has axially directly connected compression and power cylinders. A compression-side and a power-side rotational valve are separately fitted in two recessed end surfaces of the intermediate cylinder wall. The compression-side and the power-side rotational valve are provided with three L-shaped first and second openings, respectively. The compressed air-fuel mixture in the compression cylinder flows through the L-shaped first openings into the combustion chamber, and high temperature high pressure gas generated after explosion in the combustion chamber unit flows through the L-shaped second openings into the power cylinder.

ANNEX 7

United States Patent	**10,724,493.**
Choi , et al.	**July 28, 2020**

Device for increasing torque of crankshaft pulley

Abstract

A device for increasing torque of a crankshaft pulley includes: a pulley surrounding a crankshaft and connected to ***engine*** accessories; a planetary gear set having first, second, and third ***rotary*** elements, wherein the ***rotary*** element is always fixed to the pulley, the second ***rotary*** element is always fixed to the crankshaft, and the third ***rotary*** element is operated as a selective fixed element; a one-way clutch disposed between the first rotary element and the second ***rotary*** element to transmit rotational force only in a direction from the first ***rotary*** element to the second ***rotary*** element; and a first bearing disposed between the crankshaft and the first ***rotary*** element.

ANNEX 8

United States Patent	**10,598,087.**
Jeng , et al.	**March 24, 2020**

Intake/outlet pipe optimisation method for rotary engine

Abstract

An intake/outlet pipe optimisation method for a ***rotary engine,*** comprising the steps of: (A) providing a ***rotary engine;*** (B) providing a simulation software package, to perform a series of simulations for the ***rotary engine*** according to different combinations of a pipe length, a pipe diameter, a pipe shape and a pipe angle, to determine an optimal combination of the pipe length, the pipe diameter, the pipe shape, and the pipe angle, to obtain an optimal power output for the ***rotary engine***; and (C) performing tests for the ***rotary engine***, by utilizing the optimal combination of the pipe length, the pipe diameter, the pipe shape, and the pipe angle obtained in step (B), to obtain a test optimised power output for the ***rotary engine***.

ANNEX 9

United States Patent	**10,605,084.**
Gauvreau , et al.	**March 31, 2020.**

Internal combustion engine with rotor having offset peripheral surface

Abstract

A ***rotary engine*** where the rotor cavity has a peripheral inner surface having a peritrochoid configuration defined by a first eccentricity and the rotor has a peripheral outer surface having a peritrochoid inner envelope configuration defined by a second eccentricity larger than the first eccentricity. Also, a ***rotary engine*** wherein the rotor cavity has a peripheral inner surface having a peritrochoid configuration defined by an eccentricity, and a rotor with a peripheral outer surface between adjacent ones of the apex portions being inwardly offset from a peritrochoid inner envelope configuration defined by the eccentricity. The ***engine*** may have an expansion ratio with a value of at most 8. The ***rotary engine*** may be part of a compound ***engine*** system.

ANNEX 10

United States Patent	**10,612,544.**
Spencer , et al.	**April 7, 2020.**

Rotary vane device having magnet arranged in rotor and stator

Abstract

This invention relates to a ***rotary*** vane device and more particularly but not exclusively, to a ***rotary*** vane ***engine*** or pump. The invention also relates to a rotor assembly suitable for use in such a ***rotary*** vane device. The rotor assembly includes a cylindrical rotor body including a plurality of longitudinally extending receiving slots, the cylindrical rotor body further including a hollow core located radially inwardly of the receiving slots; and a plurality of vanes, with each vane being slidingly locatable inside a receiving slot. The rotor assembly is characterised in that the vanes are biased away from the cylindrical rotor by way of a magnet arrangement including vane magnets located in the vanes, and opposing rotor magnets located inside the hollow core of the rotor body.

ANNEX 11

United States Patent **10,612,558.**
<u>De Gaillard , et al.</u> **<u>April 7, 2020</u>**

Rotary assembly of an aeronautical turbomachine comprising an add-on fan blade platform

Abstract

A ***rotary*** assembly of an aviation turbine ***engine,*** includes a fan disk having at least one tooth and at least one platform mounted on the tooth of the fan disk. The tooth of the fan disk includes a tab extending the tooth axially upstream, and the platform includes a locking ring at its upstream end for receiving the tab of the tooth of the fan disk. The assembly further includes a spacer positioned inside the locking ring so as to block the platform on the tab of the tooth of the fan disk.

ANNEX 12

United States Patent **10,677,498.**
<u>Longsworth</u> **<u>June 9, 2020</u>**

Brayton cycle engine with high displacement rate and low vibration

Abstract

To provide refrigeration below 200 K, a Brayton cycle ***engine*** contains a light reciprocating piston. The refrigerator includes a compressor, a gas-balanced reciprocating ***engine*** having a cold ***rotary*** valve, a counterflow heat exchanger, a gas storage volume with valves that can adjust system pressures, a variable speed ***engine*** and a control system that controls gas pressure, ***engine*** speed, and the speed of the piston. The ***engine*** is connected to a load such as a cryopanel, for pumping water vapour, through insulated transfer lines.

ANNEX 13

United States Patent **10,697,365.**
Thomassin , et al. **June 30, 2020**

Rotary internal combustion engine with pilot subchamber

Abstract

A ***rotary engine*** having an insert in a peripheral wall of the stator body, the insert being made of a material having a greater heat resistance than that of the peripheral wall, having a subchamber defined therein and having an inner surface, the subchamber communicating with the cavity through at least one opening defined in the inner surface and having a shape forming a reduced cross-section adjacent to the opening, a pilot fuel injector having a tip received in the subchamber, an ignition element having a tip received in the subchamber, and a main fuel injector extending through the stator body and having a tip communicating with the cavity at a location spaced apart from the insert. The subchamber has a volume corresponding to from 5% to 25% of a sum of the minimum volume and the volume of the subchamber. A method of injecting heavy fuel into a Wankel ***engine*** is also discussed.

ANNEX 14

United States Patent **10,683,813.**
Sano , et al. **June 16, 2020**

Rotary-type throttling device for internal combustion engine

Abstract

A *rotary-type* throttling device for an internal combustion ***engine*** includes an upstream auxiliary intake passageway formed in a throttle body and having an inlet port held in fluid communication with the atmosphere, and a downstream auxiliary intake passageway formed in a cylindrical valve body of a ***rotary*** valve and having an outlet port open at a downstream outer circumferential surface of the cylindrical valve body. The upstream auxiliary intake passageway and the downstream auxiliary intake passageway have a body-side joint fluid communication port and a valve-side joint fluid communication port formed in respective sliding surfaces of the throttle body and the cylindrical valve body and designed to overlap each other to keep the upstream and downstream auxiliary intake passageways and in fluid communication with each other. When the ***rotary*** valve is open, a main intake air stream passing through an intake passageway in the ***rotary*** valve flows smoothly for enhanced intake performance without being disturbed by an auxiliary intake air stream flowing out of the outlet port of an auxiliary intake passage.

ANNEX 15

United States Patent **10,626,793.**
Radocaj **April 21, 2020**

Internal pressure driven engine

Abstract

An internal combustion ***engine*** or other internal pressure driven ***engine*** of the type capable of converting reciprocal linear powered motion into unidirectional ***rotary*** motion, the ***engine*** having at least one pair of first and second cylinders with each cylinder having a pair of opposed pistons therein forming a pressure chamber therebetween. The outer ends of each piston carries a piston rod connected to a pivot arm of a respective one way clutch which causes the clutch to oscillate back and forth when the piston moves in and out due to pressure or combustion in the pressure chamber. Alternatively, the piston rods may be configured as gear racks in direct operative engagement with pinion gears of the one way clutches. The clutches are parallel and spaced apart from each other near each end of the cylinders. Each clutch carries a gear on one end which intermeshes with a gear rack assembly having gears and a gear rack which drives a crankshaft and auxiliary flywheel operatively connected to a starter. Once the starter is turned on the kinetic energy of the flywheel and gear rack keeps the on/off clutches in continuous oscillation. The oscillating clutches turn unidirectional drive shafts connected through pinion gears to a main output shaft and main flywheel.

ANNEX 16

United States Patent **10,669,038.**
Scannell , et al. **June 2, 2020.**

Engine inlet system with integral firewall seal

Abstract

One embodiment is an inlet plenum assembly for a ***rotary*** aircraft that includes an inlet plenum defined on a first side by an inlet plenum wall and on a second side by a forward firewall assembly, the forward firewall assembly having an inlet aperture configured to receive a drive shaft to be rotatably coupled to an ***engine;*** and a seal connected to the forward firewall assembly via a seal retainer disposed around a periphery of the inlet aperture. The forward firewall assembly may further include a forward firewall upper portion and a forward firewall lower portion, the forward firewall upper portion configured to removably seat to the forward firewall lower portion. The seal retainer may be attached to the forward firewall assembly via a plurality of bolts disposed around the periphery of the seal retainer. In some embodiments, the seal is connected to an inlet duct of the ***engine***.

ANNEX 17

United States Patent	**10,730,508.**
Hoshino , et al.	**August 4, 2020**

Hybrid vehicle and control method for hybrid vehicle

Abstract

A vehicle is equipped with a torque sensor configured to detect a torque, and which is disposed between an internal combustion ***engine*** and a first switching device on a first transmission path. When the first switching device is switched from a disconnected state to a connected state, a control device calculates a torque difference between a target torque of the internal combustion ***engine*** corresponding to a specified point on the first transmission path, and a detected torque or the like detected by the torque sensor. Further, the control device generates in a first ***rotary*** electric machine or a second ***rotary*** electric machine a compensation torque that compensates for the torque difference.

Printed by Books on Demand GmbH, Norderstedt / Germany